SOLDIER COLORING BOOK

American Military from 1780 to Modern Times

ARTHUR BENJAMIN

This page intentionally left blank.

ABOUT THE BOOK

Explore the development of the American military uniform through this unique coloring book. Starting with 1860 and ending with modern times, watch how the US military uniform as well as their main weapons developed since the very inception of US military.

Printed in the United States of America
ISBN: 978-1619495418

CONTENTS

This page intentionally left blank.

1780

Plate 1.

1812

Plate 2.

1840

Plate 3.

1860

Plate 4.

1890

Plate 5.

1910

Plate 6.

1930

Plate 7.

1965

Plate 8.

ABOUT THE BOOK

Explore the development of the American military uniform through this unique coloring book. Starting with 1860 and ending with modern times, watch how the US military uniform as well as their main weapons developed since the very inception of US military.

This page intentionally left blank.